MW01673458

MOVING BEYOND Duck, Cover and Hold On

Emergency Preparedness Guidebook for School Volunteers, Parents and PTAs

ELLEN RONY

Illustrations by Mel Warenback

Moving Beyond Duck, Cover and Hold On

Cover art and illustrations by Mel Warenback

Published by Alexander Works
Tiburon, California

Printed in the United States of America.

ISBN-13: 978-0615753690
ISBN-10: 0615753698

An e-book version is also available in multiple electronic formats.

All web links in this guidebook are correct as of the publication date but may have become inactive or otherwise modified since that time.

Table of Contents

Acknowledgments v
Preface vii
1. Introduction 1
2. Join the Volunteer Disasterati! 3
3. Preparedness Test 5
4. Getting Started 8
5. Parent Notification 12
6. Grab-n-Go Classroom Backpacks 17
7. Student Comfort Kits 21
8. Pocket Contact Card 24
9. Storage and Supplies Cache 29
10. Parent Responsibilities 39
11. Drills and Other Activities 42
12. Resources for Families 45
13. Fundraising 54
14. Maintaining Momentum 56
15. Disaster Mythbusters 60
16. Frequently Asked Questions 64
Appendices 67
A. Acronyms 67
B. Emergency Supplies Checklist 70
C. Supplies Replacement Form 72
D. Classroom Lockdown Kit 73
About the Author 75

Acknowledgments

God could not be everywhere, and therefore He made mothers.

This guidebook is intentionally limited in size and scope, but such simplicity is deceptive. Many individuals contributed to shaping this manuscript into its final form.

First, a special gratitude to the dedicated moms who joined me for three years to develop and implement a school district emergency support program that, a dozen years later, helped shape this guidebook: Mary Barnes, Suzanne Pressley and Katie Ai. Working as a team, we developed a framework that is used in the school district to this day, and we had fun doing so. They endowed me with the moniker of Disaster Queen, but they are the dedicated Princesses of Preparedness.

My thanks, too, to the members of the Marin County Office of Education Parents Disaster Advisory Council, especially those who continue to participate year after year. You know who you are. I have learned so much from your commitment and support for emergency preparedness in Marin County schools.

I am very grateful to several individuals with hands-on experience who greatly contributed to my knowledge and expertise. Although this guidebook was not written within the scope of my employment at the Marin County Office of Education, I am indebted to Mary Jane Burke, Marin County Superintendent of Schools, and Luke McCann, Deputy Superintendent, who recognized my passion for emergency management and gave me the opportunity to engage on a professional level. Sincere thanks also to Robert Barnecut, Program Manager (retired) for his encouragement and careful reading of the draft. His practical advice greatly improved this guidebook, particularly cautioning me to avoid the Latinate verbiage with which I am so enamored.

Ursula Hanks, Emergency Coordinator of the Marin County Office of Emergency Services, has greatly enriched my knowledge in this area as a colleague and a friend. Ursula provided a thoughtful review of the draft that helped me avoid embarrassing errors. My appreciation also goes out to my long-time friend Jean Armand for pointing me toward articulating my school preparedness experience in guidebook form.

I am deeply grateful to Mel Warenback, an imaginative and talented graphic artist and wildly creative octogenarian. Mel's enthusiastic support when I first approached him with the idea propelled me through the writing. His endearing sketches impart a light-hearted touch to the serious subject of this guidebook, and I would work with him again in a heartbeat.

Constance McCarney provided valuable assistance by giving the manuscript its final read. She added stylistic consistency and showed me that even a guidebook as small as this one can greatly benefit from the expertise of a copy editor with a skillful eye.

Finally, I have abiding gratitude for the ongoing support of my family. My brother, Peter, shares my visceral interest in emergency preparedness, and he has offered encouragement throughout my 15-year preoccupation with this topic. He has greeted my boring but practical holiday gifts (headlamps, mylar blankets, solar/crank emergency radio) with enthusiasm and appreciation.

My beloved son, Alexander, without whom I never would have explored school emergency preparedness, has often heard me justify my creative projects with the explanation that woman cannot live by couscous alone. When I first floated the concept for this guidebook a decade ago, he immediately contributed his special brand of youthful levity to the task. He proposed disaster scenarios involving disgruntled mongeese and stampeding wildebeests, which he noted would be of greater concern to schools in Africa than shark attacks. Perhaps this notion was inspired by the fact that our home is in the bayside town of Tiburon, which is the Spanish word for shark.

Preface

Every day cannot be a feast of lanterns.

How many of us experienced separation anxiety when we first sent our young children off to school? Though the initial apprehension will dissipate over time, concerns are likely to persist about what might happen to our precious offspring if a crisis occurs during school time.

We live in an age of anxiety. Widely publicized accounts of tragedy striking school campuses have raised the level of concern among parents, teachers and students alike. The unsettling events at the Sandy Hook Elementary School in Connecticut, Columbine High School in Colorado and the Virginia Polytechnic Institute and State University have forced our schools to shed their innocence, dispel their complacency about school preparedness, pause and take stock. The crucial lesson drawn from those headlines is that schools must maintain a high level of preparedness for man-made emergencies as well as natural disasters. Fire drills, the long-established staple of school emergency preparation, are only the edge of empowerment in an emergency. The disasters that may beset a school today are complex and require substantially more detailed contingency planning.

Such planning is the responsibility of the school district and site administrators, but preparedness programs also depend upon support from the parent community. While policy and procedural decisions are clearly the province of school administrators and risk managers, some activities can be coordinated by parent volunteers to enhance overall school emergency planning, coordination and response. Administrators and parent organizations can work together to establish emergency preparedness activities that boost confidence in the response capability of the school community.

In the early 1990s, a small team of PTA members designed a program of activities to enhance emergency planning for 1,200 kindergarten through eighth-grade students disbursed among three school campuses in a Marin County, California community. The effort became a model for other schools in the county and other regions.

This guidebook condenses that three-year effort and the work that has followed in the subsequent decade. It was written to launch the process of organizing similar parent-led projects in other school communities. The information presented here does not represent the official policy or practices of any school or organization.

Each chapter outlines an activity that can be adapted for any school, no matter where situated, no matter the size. The emphasis is upon several prescribed projects where parent committees can take the lead to support school preparedness.

Anyone who spends time thinking about school emergency preparedness can offer a disaster scenario, a situation hopefully remote but theoretically possible. The school-related scenarios presented in this guidebook (in *italics*) describe actual incidents and introduce the parent-led activity from which they might benefit. Given the author's California residency, many of the examples, and, indeed, the title, are colored by a West Coast exposure. None, however, are doomsday scenes involving suicidal pilots, cities blasted into oblivion or tons of concrete instantly vaporized into an enduring lethal haze. All felt dangers to be certain, but this guidebook does not envision disturbing scenes of widespread human panic and loss. Nor does it contemplate building bunkers on campus or developing biometric IDs for the student body.

Despite the temptation to emphasize such heart-tugging, high-stakes perils in this post 9/11 age, this text focuses on more localized emergencies. It is targeted for schools ready to recognize that emergency preparedness no longer consists solely of monthly fire drills, in which students evacuate classrooms to a pre-designated spot, or a quarterly duck, cover and hold on exercise to prepare for when the earth beneath them starts to rock and roll. This guidebook is designed to jumpstart the process toward supporting the resiliency of schools and families to respond to emergencies.

There is no ironclad way to prepare for disasters in the contemporary world and no sure means to neutralize our vulnerability, but a thoughtful approach that addresses foreseeable communication, logistical and safety issues can enhance our peace of mind while our beloved offspring are away at school.

Introduction

Denial is not just a river in Egypt.

May in the Southwest is gentle, if dry. In summertime, temperatures often soar into the high 90s, and wildfire alerts are commonplace. By September, vegetation languishes from dehydration, and the rolling hills and urban forests turn brown and brittle from months without moisture.

The heavily wooded hills of Oakland, California were combustible fuel for a firestorm in October 1991. According to eyewitnesses, a single ember from a fire being contained from the previous day blew into a tree just outside the burn zone. The tree instantly exploded into flame. Fierce winds propelled the blaze as it gusted through flammable groves of freeze-damaged eucalyptus trees, Monterey pines and underbrush parched from five years of drought.

The winding, narrow streets were immediately clogged with residents trying to leave and sightseers and emergency personnel attempting to get in. Steep terrain, heavy chaparral, downed power lines and utility poles, smoke and ash–coupled with record high temperatures and the rapid sweep of the firewall–made movement of firefighting apparatus impossible. Houses vaporized instantly, and according to eyewitnesses, a "big sheet of fire rolled over the area like a wave."

The inferno advanced destructively in many different directions, overwhelming the fire suppression efforts. It moved with such velocity that it tore trees apart. Directly in its path was Kaiser Elementary School, a magnet school for approximately 185 kindergarten through sixth grade students. If there were ever any blessing to be found in the terrible disaster that befell Oakland that day, it was that the firestorm ignited when Kaiser School was not in session. There were no students to evacuate, no fire response by the faculty, no teachers to assign to medical care.

The conflagration was one of the nation's most costly fires. Trees on the hill behind Kaiser Elementary School and houses across the street burned to the ground. The wooden play structure and a ring of trees on the property also burned. Remarkably, the buildings were spared of fire damage because the blacktop parking lot that surrounded the school provided an essential firebreak. With the incendiary event, residents in the region shed their innocence that day, but a Kaiser School teacher proclaimed, "The goddesses were looking out for us."

Disaster strikes in highlands and canyons, greenbelts and suburbs, along coastlines and deep within heavily populated metropolitan districts. Every geographic region endures its own special type of natural vulnerability—blizzards, floods, earthquakes, mudslides and landslides, tornadoes, tsunamis, hurricanes, drought, wildfires, windstorms, land insect infestation. Preparedness efforts have typically addressed these naturally recurring catastrophes. But one of the lessons learned from the 9/11 terrorist attacks is that no disaster scenario is too improbable, illogical or grim. Preparedness must now include emergencies inflicted by human activities–bomb threats, chemical spills, intruders, gang violence and more.

Proper preparation is critical to the successful management of threatening situations on school campuses. Specific situations require specific responses. One school may monitor the uninvited presence of an errant coyote, while another has prophylactic measures in place in the event of a chemical spill at a nearby processing plant. Our children and schools benefit when parents learn how to prepare for emergencies–what precautions to take, what school procedures are in place and how to respond. An imperative we can take from the school emergencies that have made headline news is that they obliged schools, PTAs and parent councils to reflect on school planning and implement a preparedness program like never before.

Join the Volunteer Disasterati!

Good luck happens when preparedness meets opportunity.

Top Ten Reasons to Volunteer

10. Avoid all the marketing calls you get at home.
9. Put your higher education to good use.
8. Your kid might need help some day.
7. Earthquakes. Wildfires. Rabid animals. Ex-spouses.
6. Scenario planning is a hoot.
5. Afternoon soap operas are all alike.
4. Why should teachers have all the fun?
3. No one ever understands disaster until it occurs.
2. Noah's Ark was built by amateurs; the *Titanic*, by professionals.
1. Survive the *whangdepootenawah* (the Ojibwa word for disaster, an unexpected affliction that strikes hard).

What skills are required of those offering to take the lead in designing and maintaining an emergency preparedness support program for schools? Objectives here are impossible to glamorize, but a passionate parent usually can overcome the apathy factor that often characterizes this sphere of activity. Parent volunteers should be resourceful, resilient, fearless and curious. They must possess a steadfast commitment to the task. An ability to do scenario planning is another requisite. That means, simply, the parent team must know what tough questions to ask and where to go for answers. The key, however, is enthusiasm for the project.

Volunteers may be hard to identify when parents assume that emergencies are unlikely events or effective planning is not possible. Some parents downplay the risk of disaster as cognitive protection against the daily uncertainty of modern society. They may choose not to read the media churn about how to cope with threats of imminent disaster or think about potential danger that may accompany the outbreak of a new flu virus in classrooms. It may be easier to consider "contingency planning," rather than emergency preparedness to describe the unpredictable, unplanned and uncontrollable events that bring injury or death to the school community, disrupt classroom activities and damage school facilities.

Preparedness Test

A cracked bell can never sound well.

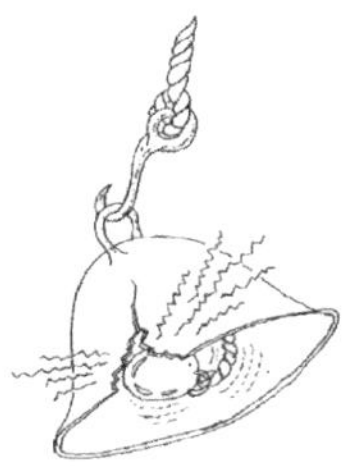

This is an exam to test the emergency awareness of parents.

ALERT: This exam contains humorous flourishes to expose common myths and misconceptions.

School Name: ________________________________

Number of students: ___

Number of teachers who commute more than an hour: __

Proximity to the nearest Starbucks: ________________

Teachers' preferred brand of athletic shoe: ___________

Please check all that apply:

- ❑ Twisters only happen in the Midwest.
- ❑ Earthquakes only happen in California, Pakistan, Chile and Indonesia.
- ❑ Fire drills prevent fires.
- ❑ There is always advance warning for storms.

In an emergency, parents with children in schools may be (check all that apply):

- ❑ anxious
- ❑ demanding
- ❑ hysterical
- ❑ impatient
- ❑ irrational

In the event of an earthquake, you should (check all that apply):

- ❑ duck, cover, hold on and rock and roll
- ❑ use your cell phone to call all loved ones
- ❑ create a cell phone video for YouTube
- ❑ apply makeup in advance of a visit by the local TV news crew

Which emergency is most commonplace?

- ❑ rabid bats
- ❑ rabid spouses
- ❑ broken water pipes
- ❑ electrical storm
- ❑ sold-out tickets for a Rihanna concert

Hemorrhage is usually associated with which of these (check all that apply):

- ❑ wounds
- ❑ choking
- ❑ teaching your teen how to drive
- ❑ dislocations
- ❑ food poisoning

Emergency preparedness at school means (check all that apply):

- ❑ evacuating without having to say you're sorry
- ❑ monthly five-minute drills
- ❑ handicap access to the lunchroom
- ❑ drinking water, MREs (Meals Ready to Eat) and space blankets
- ❑ cinnamon dolce latte in hand; 25 students in single file
- ❑ school is closed for the day

When there is an emergency at school, you should:

- ❑ rush to the campus to pick up your child

- ❑ telephone the school office to get the details
- ❑ drive to the supermarket to purchase batteries, water and food

The best reason for thinking that the worst won't happen is because:

- ❑ it has not happened before
- ❑ it has not happened in your lifetime
- ❑ people of the preparedness persuasion tend to exaggerate

Getting Started

Even the loftiest mountain begins on the ground.

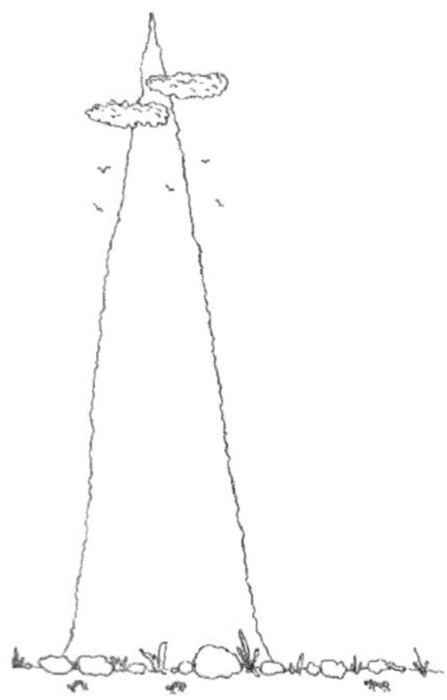

At 4:42 p.m., an earthquake registering 6.5 on the Richter scale struck with a sharp jolt. The aftershock seconds later registered at 5.0, causing widespread destruction and hindering response efforts.

Windows imploded, littering the rooms with daggers of glass and gouging wooden furniture with flying shards. Filing cabinets and bookcases skated across floor tile, crashing into desks and blocking exits. Cabinets sprung open and spilled their contents. Books, laptops and other school materials became temporarily airborne. Florescent bulbs dislodged from their fixtures and shattered on the floor. Water pipes broke, too, flooding basements and destroying electrical circuitry. Fortunately, school was not in session, but the earthquake rendered the building structurally unsafe.

Emergency planning begins with a question:

WHAT IF?

WHAT IF the community is devastated by an earthquake, tornado, blizzard or hurricane?

WHAT IF the school is in the path of a wildfire?

WHAT IF the National Weather Service issues a flood warning for low-lying areas near small streams, including an elementary school?

WHAT IF a rabid animal wanders on to the school grounds?

WHAT IF there is an airborne release of hazardous chemicals nearby?

WHAT IF an angry teen threatens students and staff with harm or a school intruder takes hostages?

WHAT IF fallen trees and electrical wires block parents from getting to the school to pick up their children?

Each of these "what ifs" requires scenario planning that will set the direction for the appropriate response. The carefully detailed answers to these questions provide a framework for the policies, procedures and planning that must be established for the school community. What is the scope of the task? Someone has to do it, but who?

In most cases, PTA committees or parent groups take the lead in the peripheral areas while school administrators develop the policy guidelines and procedures. The partnership between school staff and parents working together on emergency planning makes effective response far easier to achieve.

PTAs and parent organizations provide volunteer assistance under the direction of school district officials, site principals or their designees to support the school, staff and students' preparedness activities for any type of emergency. All it takes to get started is an enthusiastic parent who is interested in being actively involved. However, working as a committee is definitely more fun and prevents the activity from becoming a desultory affair. At many schools, this committee oversees the assignment of the parent telephone tree, if one is needed; the purchase and storing of emergency food, water and first aid supplies; and the collection of student comfort kits for elementary school students. The parent-led committee may also provide resources that offer guidance in home preparedness for families and may be asked to assist the school with its evacuation exercises and student release drills.

The committee chair should be prepared to cooperate with the school principal or staff emergency preparedness coordinator and provide periodic updates to the parent organization, school site council and, possibly, to the school board. Many of the activities require little investment of time or money, just a conviction that emergency empowerment is both a necessary and a realistic goal.

The California State PTA provides a job description for a Disaster Preparedness/Crisis Response Committee, summarized in the Maintaining Momentum chapter of this guidebook.

A parent-led Disaster Preparedness/Crisis Response Committee might include the following activities:

> ENCOURAGE participation in the annual review of the school's safety and emergency management plans and share the plans with parents and the community.
>
> REVIEW the American Red Cross, Federal Emergency Management Agency and PTA publications that are free for distribution.

ESTABLISH an ongoing working group with the school district/site and safety agencies within the community (e.g., local branches of the National Safe Kids Coalition and the National Safe Routes to Schools Partnership).

ATTEND safety workshops, conferences and clinics. Cooperate with school authorities to inform the parent community of these programs.

PLAN a safety/preparedness program for at least one meeting during the year.

SPONSOR parent education safety awareness programs.

The remaining chapters outline activities that can be developed and maintained by parents to support the emergency preparedness program of a school site:

- Parent Notification
- Grab-n-Go Classroom Backpacks
- Student Comfort Kits
- Pocket Emergency Card
- Storage and Supplies Cache
- Resources for Families

What if the school administrator is reluctant to admit that the site preparedness is not up to par and is resistant to volunteer help? This is not a zero-sum game. Everyone benefits when parents work with teachers and staff to enhance disaster protection for the students and raise the standard of preparedness in the school.

Avoid the notion that these tasks are too expensive, expansive or impermanent. Each step that the volunteer group takes diminishes the potential for chaos and concern during an emergency.

Parent Notification

Rain does not fall on one roof alone.

One of those rare polar masses blew in from Alaska. It rained throughout the night, and sleep was fitful as the wind howled and gusted through the trees. Twice during the night, the woman arose to check on her 8-year-old son. She awoke at 6:45 a.m. to the sound of heavy raindrops and the insistent ring of a telephone a few inches from her head. In a sleep-deprived stupor, she dragged her arm over to pick up the receiver. The voice on the other end was alert. "The telephone tree has been activated."

The phone message was short and as crisp as the one delivering it. "School will be closed today due to the storm. Please advise the parents on your list immediately. Updated information will be available on the school's recorded hotline."

Now she was bolt upright and fully awake. She wanted to know what other schools were closed. Had the storm pushed floodwaters gushing through the canyons? Had it destabilized the land, causing homes to slide toward more solid ground? But the caller was already gone. She reached in the drawer of her nightstand and pulled out her telephone tree list. Only five calls to make below her name.

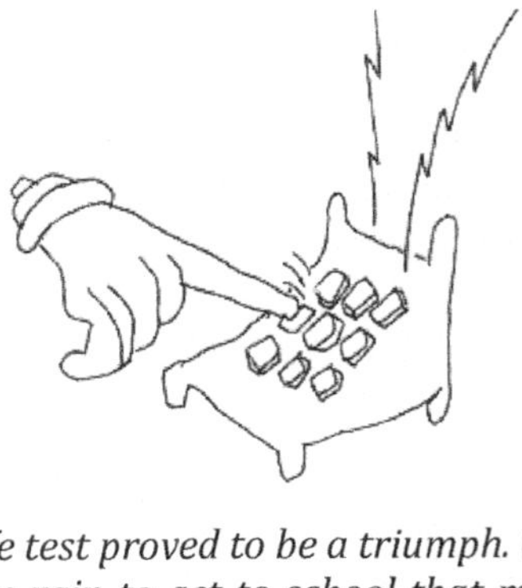

The telephone message went out in a distributed hierarchy to more than 600 families within 45 minutes. By the time the school community had been contacted, the storm had passed, leaving uprooted trees and impassible roads in its wake. For those who had designed the parent communication system, this real life test proved to be a triumph. Only four students had slogged through the rain to get to school that morning, unaware that the campus was closed.

The telephone tree provides a way to notify all parents of a developing emergency involving a school. Increasingly, however, schools are establishing automated telephone notification systems that can reach hundreds of cell phone and landline numbers very quickly with a brief recorded message. In the nomenclature of emergency responders, this is also known as TARP (Telephone Access Routing Plan) or TENS (Telephone Emergency Notification System).

Some systems encourage parents to go online and insert their home, work and mobile phone numbers, and the automated message will reach out to all three. There are also automated notification systems that will send out email alerts.

Such systems are pricey, and not all school districts can afford them. The telephone tree can be an effective alternative when properly designed and used. It allows an authoritative source to convey reliable information at any time of the day or night. It includes all the school families and is normally activated through the PTA or parent group structure upon notification by the superintendent or school principal.

The structure is designed to reduce the volume of calls made by a single individual. No parent should be asked to make a dozen calls during a developing emergency. Establishing the annual telephone tree assignments and explaining how the process works in an emergency should be included among the activities of the parent volunteer emergency preparedness committee.

The common way to distribute this communication task is to mirror the structure of the district PTA or parent organization. When the telephone tree is activated, the individual at the top of the hierarchy, generally the PTA/parent group president, generally calls members of the Executive Committee. In many cases the schools within a district will each have a PTA vice-president/parent chair. They in turn, might phone the coordinators for each grade at the school. The grade coordinators will call the room parents designated for each classroom and they will phone parents of the students in the class.

In the ideal telephone tree activation, that is how it works. The communication hierarchy is distributed throughout the school. However, parents cannot always be reached by telephone or cell phone. The calls may go unanswered or may not be heard. The cell phone may not be charged. Or a recorded message may greet the caller.

If the caller is unsuccessful in reaching someone who is expected to phone others, the caller must make those calls, too, or find someone else who can complete the task so that the tree chain is not broken. Rapid response and teamwork are essential to the success of the system.

Telephone Tree Basics

- Establish the telephone tree at the beginning of the school year.
- Assign each person no more than five phone numbers to call.
- Keep copies of the phone numbers in your nightstand, wallet and car.
- When you receive a call, phone everyone on your list immediately.
- Remember KISS. Keep the telephone message short and simple.
- Stick to the script! Do not engage in discussions and conversations.
- Remind parents to call the designated hotline at school for emergency updates.
- Do not assume that messages left on voice mail will be received. Call back.
- If you have more than one child in the school, expect duplicate messages.
- Be calm.

When the telephone alert at the beginning of this chapter was completed, a parent survey was distributed to review the effectiveness of the telephone tree. The survey identified areas where people were unclear about its purpose and limitations.

Frequently Asked Questions About Telephone Trees

Q. How do we sign up for the telephone tree?

A. Every family at every school is automatically included. No one needs to sign up, but it is essential to keep the telephone information current.

Q. Where is our family on the telephone tree?

A. The telephone tree provides a communication hierarchy that usually mirrors the organizational structure of the PTA or parent group through the grade levels at the school sites.

Q. Why weren't we called before we had to leave for work?

A. Emergencies rarely occur at convenient times. The telephone tree is activated as soon as a need for it is determined. Families are phoned as quickly as possible.

Q. Our family received calls for both of our children. Can this duplication be avoided?

A. This built-in redundancy is more efficient than assuming that parents have already been notified through other branches of the tree. If the system operates correctly, families with more than one child in the school should receive a call for each child in a different classroom.

Q. The message was too brief. Why weren't we given more information?

A. An emergency is a rapidly changing situation. By keeping the information brief, the message will be transmitted to many families quickly and accurately. The call should alert parents to turn on their radios or call the school recorded line for updated information.

Q. Our telephone did not work, so no one called us. How do we find out what is happening?

A. Since smoke signals and runners aren't realistic alternatives, concerned parents should call the school's recorded hotline or listen to the local radio station. In the event of a power outage, cordless phones won't work and cell phone transmitters may be overwhelmed in an emergency. Keep a standard landline phone connected to one outlet so that you retain phone service even if you lose electrical power.

Q. Why wasn't my ex-spouse called?

A. Calls are made to the main residence of each student according to the primary telephone number listed on the emergency contact card. Other caretakers and relatives are not generally called when the telephone tree is implemented.

Q. Why does it take the radio station so long to list school closings?

A. Radio stations must announce traffic advisories, accidents, bridge access, as well as school closings. That is why a telephone tree is a more direct and effective way to transmit information to the school community.

Q. What should we do about people we cannot reach? Their phones went unanswered.

A. Leave a voicemail and continue calling your list. But do not assume the voicemail message will get picked up immediately. If you are unable to contact someone who is expected to phone others, make those calls, too.

Q. Does the telephone tree change every year?

A. Yes. The telephone tree numbers must be reassigned annually as new families enter the school and older students graduate. In addition, the volunteers who hold positions within the PTA/parent organization change from year-to-year, but the telephone tree hierarchy should remain the same unless the underlying organizational structure has changed.

Grab-n-Go Classroom Backpacks

Nothing makes a person more productive than the last minute.

The spring afternoon sparkled with a breathtaking brilliance for which the area is renowned. Young students sat in their classrooms practicing simple arithmetic and reading about the antics of Curious George. Outside, the happy chatter of high-spirited second graders playing dodge ball mixed with the fragrance of wild honeysuckle and the screech of a distant blue jay.

On a hill behind the school property, contractors cleared overgrown trees with large electrical equipment. In the process, an unseen water line was broken and a cascade of water burst from the pipe, drawn by gravity to lower ground. Teachers quickly gathered their young charges, counted heads, grabbed the red emergency backpack hung by the classroom door, and evacuated the school single file down to the main street of town. When they were a safe distance from the threat of immersion, the teachers gave a pre-printed name label to each student. Grateful for the springtime warmth, the teachers again assured that every child was accounted for.

The local fire and police departments were on the scene immediately, but the shut-off valve was buried beneath concrete boulders and a decade-old thicket of shrubs. Within minutes, the water was turned off at an upstream valve. After the damage was assessed–none as it turned out–and it was deemed safe to return, the children were brought back to their classrooms. The break in the school routine kept everyone abuzz for the rest of the day. Only a minor emergency, but what an adventure for these five-and six-year olds! What exciting stories they would bring home to their families that afternoon.

The full magnitude of a localized emergency is rarely known immediately. It may be over in a matter of minutes or it may overwhelm the school for months. When it is time to assemble the students and respond to the emergency, teachers do not want to be scrambling through desk drawers in search of the procedures and class roster. The Grab-n-Go classroom backpack, a small emergency carryall mounted near each classroom exit, facilitates immediate use by the teaching staff. What it contains depends upon the age of the students, the setting of the school and other considerations. It should be lightweight and include the minimum materials needed for a short-term evacuation or lockdown. The classroom backpack serves both types of responses.

In an evacuation, the teacher leaves the classroom with all the students and may not return for the remainder of the day. As a minimum, teachers need a classroom list, a whistle to get the attention of students, pre-printed name labels and red, yellow and green sheets to indicate to the Incident Commander the status of students in their charge when the evacuation accountability assessment is taken.

In a lockdown, as might be required if there is a report of a rabid animal or intruder on campus, the teacher and students are required to stay inside the classroom with the door locked until the emergency is resolved. A simple snack to share with students is welcome in such circumstances.

Now come the WHO, WHERE and WHY.

WHO should have a backpack? The answer is that it is more appropriate to ask where it should be placed, rather than who needs one. A Grab-n-Go backpack should be hung next to the emergency exit of every classroom and every area where children gather, even those rooms with students attending for only one period in the school day. An emergency may strike when students are in the gymnasium or multipurpose room, music room, computer room, art room or library. Each of these rooms should have its own backpack, available to the teacher using the room at the time.

WHERE should the backpacks be kept? It is essential that each backpack be accessible, not stored in closets or under desks. If evacuation is required, a teacher may not be able to open the closet door or reach under the desk to pull out the backpack. Instead, use a large red hook to mark the spot where the backpack is hung, approximately at shoulder level, near the emergency exit of the room. The evacuation route should be posted nearby.

WHY a backpack instead of a tote? It lets teachers transport the contents yet leaves their hands free for other tasks.

Classroom Backpack Basics

- Use lightweight backpacks instead of totes.
- Keep all backpacks identical and brightly colored. Red is preferred because the color is often used in association with emergency supplies.
- Mark backpacks with a red cross or other designation so that they are not confused with student packs.
- Secure the contents of the backpack with a self-locking zip tie.
- Hang the backpack near the emergency exit of each room.
- Instruct children not to touch the backpack.
- Do not store dangerous supplies (matches, medications, flares) in the backpack.
- Keep the contents lightweight for easy transport.
- Check and update the contents annually.

Parent volunteers can easily accomplish filling these backpacks if they work in partnership with the school administration. The basic classroom backpack should contain the following items:

- class list
- pre-printed self-adhesive name tags
- small first aid kit (band aids, tweezers, ice bag, gauze rolls, ace wrap, etc.)

- pre-moistened toilettes
- vinyl or non-latex (nitrile) gloves
- marking pen
- small notebook and pen
- whistle
- flashlight and batteries
- Quick Reference Guide to Emergency Actions

Optional items may include:

- scissors
- book
- mylar or space blankets
- large hefty bags (for use as raincoats or to sit on)

Teachers often personalize the contents with items such as comfortable shoes, socks, sunscreen, sunglasses, extra keys for house and car, light jacket, toiletries, work gloves and dust mask.

Self-locking zip ties in quantity can be purchased at low cost, and they will prevent unauthorized fingers from accessing the contents of the Grab-n-Go backpack. In addition, they will simplify the annual task of checking bags for items that need to be refilled.

Self-locking Zip Ties

If the emergency is likely to last several hours, school staff should access a storage container where heavier supplies such as shovels, tents, blankets, food and water are kept.

Student Comfort Kits

A hurricane derives its power from a calm center.

Traffic was snarled for more than two hours as police investigated a suspicious bag at the mass transit station in the early afternoon. The train operator had spotted a duffle bag tied to some pipes beneath one of the platforms. The local bomb squad was called in to investigate. The incident brought travel on the busy interstate highway to a standstill while the bomb squad checked the bag for explosives.

All they found were some clothes, a book and a bottle of liquor, but parents caught in the gridlock were unable to pick up their children at the end of the school day. School staff began calling alternate contacts and distributed the comfort kits to the three-dozen students temporarily stranded at school. By 5:00 p.m., all were safely on their way back home.

The Comfort Kit Program is designed to provide a short-term means of calming elementary school students while they are waiting to be picked up by a family member or authorized adult when an emergency occurs. In most circumstances, however, children will remain at school for the normal academic day.

The comfort kit is a small baggie with a few items to entertain, inform and reassure the child until his or her parents arrive. Unless strict size and content limitations are provided to parents, the comfort kits sent to school may be huge and heavy, their contents perishable. The distribution and collection of the comfort kits is an appropriate task for the Parent Emergency Preparedness Committee.

To assure that the comfort kits are not unwieldy, it is recommended that instructions be provided to each family, along with a one-quart zip-lock style transparent baggie. Otherwise, the comfort kits are likely to arrive in containers of many types–plastic, paper, metal and cloth; in paper bags, cardboard boxes, lunch pails, baskets, bowls, fish tackle boxes and even pillow cases. Supply the baggies so that parents don't pack up the pantry and send it to school!

The baggie should contain a short version of the emergency card to be completed by the parents. This identifies the student to who the bag belongs and other useful information. Advise parents to use only the small baggie provided and apply common sense when putting together the comfort kit. Liquids, glass or perishable items should not be included. The school will provide water.

Insist that parents forego homemade cookies that turn moldy, salty snacks that induce thirst, and chocolate, which may melt and make kids hyper. Parents should select an entertainment item that their child will enjoy but not one that is loud, breakable or distracting. The baggie is intentionally small so that the choice of toys is limited. There is no room for board games, building blocks, Barbie or Elmo.

A letter of assurance from the family should explain that they may be apart for a while but will be together again as soon as they are able. Parents should remind their children that it is important for their own safety and the safety of others to follow the instructions they are given while the problems caused by the school emergency are being managed. Explain how the child will be cared for, who may pick up the child and why it is important for the child to remain on the school grounds.

Two essential items are missing from this list: medication, if needed, and liquid refreshment. Student medication should be given to the school nurse with instructions for its administration. In the event of an evacuation, the nurse or a designated individual will bring those medications to the evacuation area. Drinks are heavy, and the containers can break or leak. Schools plan ahead to have water on hand stored away from the building.

Comfort Kit Contents

- ❖ Small, nutritious non-perishable snack food with a 9-month shelf life. Good choices are granola bars, dried fruit snacks and raisins. Avoid salty and homemade snacks.
- ❖ Small family photograph of parents, siblings and pets.
- ❖ Short, sealed note with words of love and comfort to reassure the child that the parent/caretaker will arrive as soon as possible. The note might also remind the child to listen to the teacher and stay with the class group.
- ❖ Small toy or source of entertainment such as a book, puzzle, playing cards, disposable camera, puppet, notebook and pen. Stick with small, lightweight, unbreakable, inexpensive items.
- ❖ Optional items: emergency space blanket, plastic rain hat or shower cap, small package of tissue.

Storage of the comfort kits must be practical and accessible. The kits cannot be evacuated with the students every time there is a drill or emergency. Duffle bags for each classroom, holding all the filled baggies, serve this purpose. Schools are advised to store the duffle bags along with other supplies in a storage container away from the building.

Teachers may hold an annual end-of-year party where the unused comfort kits are returned to the students. The students read the notes, eat the snacks and have an opportunity to celebrate an emergency-free year.

Pocket Contact Card

Today is the yesterday you worried about tomorrow.

Several children had become ill at rehearsal for a winter concert with a sudden onset of fatigue. Initially, two students began feeling faint. Their symptoms were headaches and generalized pain throughout the body. When two more students complained of dizziness and similar discomfort, the school nurse became concerned. Medical attention was advised, although not critical, and the parents of all four students were contacted to pick up their children.

The call came in to the man at 11:45 p.m. during an early working lunch. The student's mother, normally available for such unanticipated emergencies, was visiting her family out of state. Dad worked in the city, across a bridge and at least 40 minutes away when traffic was light. What could he do on such short notice? Who could he call to pick up his son? Maybe his kid's soccer coach? No, he worked in the city, as well. Perhaps the young neighbor at the end of the street who also has a son in the school? Now what was her telephone number? His wife would know, if only he could reach her by cell phone.

Wait! Didn't she give him a wallet-sized card with the names and phone numbers of those they authorized to pick up their son at school? He opened his wallet, found the pocket contact card, and phoned the first alternate listed.

What's in your wallet? Parents fill out an emergency contact card at the beginning of each school year that is kept in the office on file for each student. The card records all telephone numbers and addresses where parents, relatives or three friends may be reached so that a contact may be made as soon as possible in the event of an emergency. Most school policies prohibit releasing students to individuals not listed on the emergency card or authorized by the parents through other means.

Parents could be anywhere when an emergency strikes–at work, visiting relatives, attending a conference, shopping in another community. In the rush to complete this information, they may forget whom they have designated as alternate contacts. Or a spouse may not remember the telephone number of a child's pediatrician.

For these reasons, a small contact card that fits easily inside a wallet alongside other business cards can serve as a handy reminder.

Printed on both sides, one side should hold information specific to the child, e.g., phone numbers for the school, pediatrician and three individuals the parents have authorized to pick up the student.

The other side should include the school emergency hotline and/or local radio station to tune to for emergency information. If a parent is authorized to pick up other students, their names should be listed as well.

Ready-made wallet-size contact cards can be found online so that they do not need to be reinvented.

The Ready! website managed by the Federal Emergency Management Agency (FEMA) at READY.GOV provides a complete Family Emergency Plan that includes a template for a wallet-size emergency contact card. Rather than being double-sided, it folds in half and provides room to insert whatever other information is desired.

The FEMA Emergency Contact Card is shown on the following page.

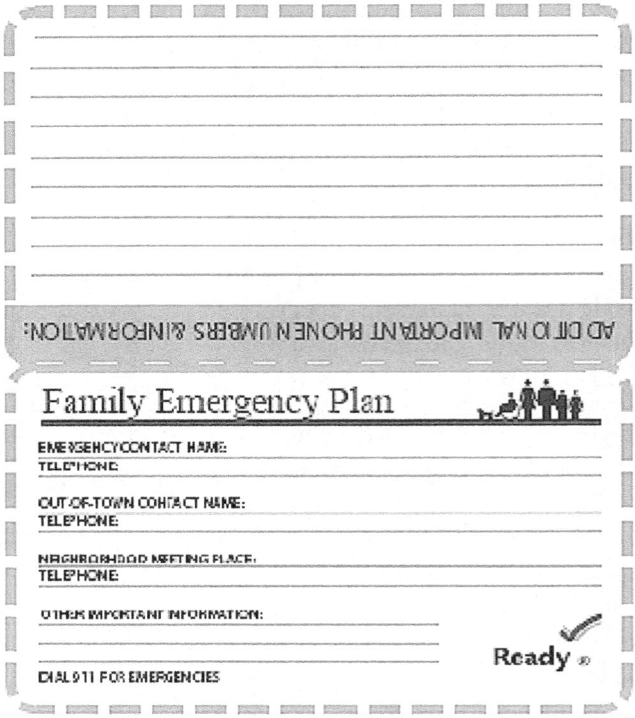

FEMA Emergency Contact Card

Source: Federal Emergency Management Agency
http://www.ready.gov

The American Red Cross has also developed an Emergency Contact Card that may be folded twice to fit inside a wallet. Alternately, using image-enhancement software, the two halves of the long card can be designed so that it is printed double-sided, providing a card that is not so bulky. The front and back sides of the American Red Cross Emergency Contact Card are shown on the following pages.

When these cards are distributed at the beginning of the year, parents should be reminded of the following:

- Parents and authorized contacts should pick up children as soon as possible.

- Children must be signed out and will only be released to a parent or person the parents/caretakers have designated on their child's School Emergency Contact Card.
- Make certain your child knows whom you have authorized for pick-up and what procedures you have chosen.
- Provide the school with accurate phone numbers to reach you in the event of an emergency. If home or work phone numbers change, or you change authorized contacts during the school year, notify the school in writing immediately.
- Call 911 ONLY if you have a life-threatening emergency. Do not call 911 solely for information.

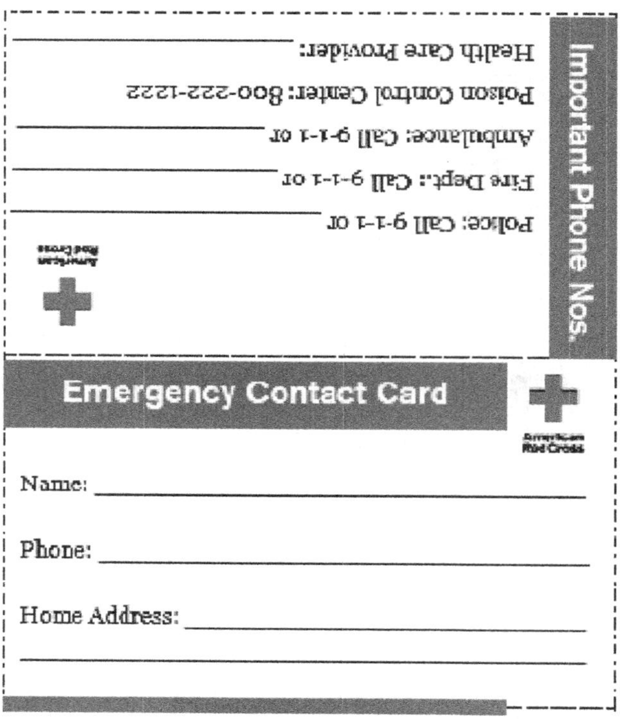

Important Phone Nos.

Health Care Provider: ______

Poison Control Center: 800-222-1222

Ambulance: Call 9-1-1 or ______

Fire Dept.: Call 9-1-1 or ______

Police: Call 9-1-1 or ______

Emergency Contact Card

Name: ______

Phone: ______

Home Address: ______

American Red Cross
Emergency Contact Card (front)

People to Call or Text in an Emergency

American Red Cross

____

____

____

____

____

____

____

____

____

____

____

Out-of-Area Contact Person: ______________

Phone: ______________________

Meeting Place Outside of Neighborhood: ______________

____

American Red Cross Emergency Contact Card (back)

Source: American Red Cross
http://www.redcross.org/

Storage and Supplies Cache

The night might be dark, but the apples have been counted.

A family in an affluent California community described the contents of their home emergency kit: Pellegrino water, powdered lobster bisque, freeze-dried heirloom tomatoes with orzo, chardonnay concentrate, battery-powered DVD player, emergency Rolodex, backup iPod to amuse 10-year-old son, list of therapists who make house calls, prepaid tickets to Cancun and a pair of LED headlamps. Don't leave home without them!

It is called a storage shed, emergency trailer, disaster cache, cargo container, or that large ugly grey thing over at the edge of the schoolyard. An integral part of emergency preparedness is stocking a storage container with equipment for search and rescue activities, first aid supplies, food and water and other items to ensure the safety and well-being of the school students and staff during and after an emergency. An earthquake, fire, hurricane, tornado or winter storm may render school buildings unsafe, making outdoor storage a sensible alternative. The objective is to have available a cache of supplies to sustain the school community in the aftermath of a disaster for up to 72 hours if the area is cut off from outside help.

The school or district will usually purchase the storage shed and choose its location. If the PTA or parent group steps up with the funding, it is best to let the school administrator determine the features desired and oversee the location, installation, maintenance and security of the shed. Rigid flooring is critical since even slight settling may cause the heavy doors of the shed to rack and bind. Many schools prefer sheds that are 8' x 20' with an 8' height clearance that allows for adequate headroom. Good ventilation is essential to maintain climate control, with intake vents on the sides and turbine vents on the roof.

The storage facility makes an attractive target for theft or vandalism, and it should not be left unlocked. A student mural painted on the most visible sides of the structure will incorporate it into the campus and make its purpose seem less threatening. Its exterior walls, however, should not be used for handball or basketball.

Purchase of the contents and their replacement and organization is an appropriate task for a Parent Emergency Preparedness Committee.

Storage Considerations

- What features do we need in a storage container?
- Where can a storage container be purchased?
- How can we assure security of its contents?
- What type of lock is best?
- What is the most practical location for the container?
- How can ants and small rodents be prevented from raiding the contents?
- How much water is enough?
- How will the water be stored?
- How will we know when the contents are out of date?
- When should food and water be replaced and supplies replenished?

Acquiring, organizing and maintaining a cache of emergency supplies are the most costly and complex activities described in this guidebook. And, perhaps, the most essential. Space and budgetary limitations and the need for strict organization of the container require that particular care be given in the purchase and placement of items in the container.

Items should be stored in sturdy, open metal shelving that is bolted to the walls of the container. If the shelves are installed upside down, they will have a one-inch ledge that serves to prevent boxes and supplies from sliding off in the event of an earthquake.

The guiding principle for organizing the storage container is LIFO (Last In, First Out). Equipment should be stored in order of use. Keep similar items together so that there are sections for search and rescue, building damage assessment and maintenance, traffic control, medical/first aid, administration, student release, food/water, blankets/clothing and entertainment.

All small supplies should be stored in sturdy, clear plastic containers with tight, snap-on lids. The contents of each container should be labeled with a laminated sheet on the inside front of the box. Colored vests should be stored for every staff member, marked with their role in emergency response.

Tape a laminated diagram of the supply container's interior near the inside entrance, with all the supply sections marked. Post in a visible spot a dated inventory of stored supplies, shelf life replacement schedule and vendor contact information. Attach a clipboard to a hook at the front of the container to maintain a wish list and missing, needed or replacement supplies.

Develop procedures to keep the container and its contents functional, clean, dry, up-to-date and in sound condition. Inspections should be made every six months. Install a maximum/minimum thermometer to determine the temperature extremes inside the container. Structural maintenance may involve resealing of exterior seams, repainting of exterior.

Every item placed in storage must address a need and satisfy other criteria, including:

- equipment that is durable and resistant to rust or mildew
- food with a long shelf life
- food that is resistant to wide temperature fluctuations
- food that requires no refrigeration, preparation and little or no water
- sufficient quantity for community served
- safe and secure location

There is no definitive supply list for school emergency preparedness. Certain items are obvious–flashlights, batteries, class rosters, gloves, ID vests, tarps, water and food, shovels, wrenches and hardhats. But when you concentrate and ask all those pesky "What if?" questions, you might remember to store traffic cones and yellow tape to reserve parking space for first responders, laminated maps to assist first responders unfamiliar with the site, shut-off list for utilities, bolt cutters, bullhorns and other less common supplies.

School Administration Documents

The California Attorney General and State Superintendent of Public Instruction's Safe Schools Task Force developed an 18-page "Crisis Response Box Guide to Help Every School Assemble the Tools and Resources Needed for a Critical Incident Response." The Crisis Response Box holds all the essential information needed for school administrators to quickly and effectively manage a major incident or emergency. A critical incident information inventory is included in the Guide and recommends: teacher/employee roster, incident command system assignments, list of students with special needs, students and staff members with allergies, daily attendance roster, emergency contacts and resources, campus maps, blueprints and aerial photos, master keys, fire alarm procedures and evacuation sites.

Assembling this critical information is not a task that will be delegated to parent volunteers, but school administrators may not be aware of this excellent resource that details the process. The time to gather this information is not when a crisis is unfolding.

Crisis Response Box

Source: California Department of Education
http://www.cde.ca.gov/

Search and Rescue Supplies

Store search and rescue supplies near the front of the container for immediate access. The PTA committee or parent club may purchase these supplies, but the school maintenance director or custodian knows best what is needed. Walkie-talkies are as important as facemasks, hardhats and heavy-duty gloves. Axes, bolt cutters, duty wrenches, shovels, sledge hammers, hack saws, crowbars and pliers are equipment needed for search and rescue. The personnel assigned to this task will also need rope, rain ponchos, school maps, keys to all rooms and whistles.

Building Assessment and Maintenance Materials

Many of the items needed for damage assessment are similar to those used for Search and Rescue. Storage for building assessment should include long, heavy-duty power cords, headlamps and flashlights, ropes, plus a generator, if the cost is not prohibitive for the budget. It is also important to include master keys and an architectural blueprint and site map of the school buildings with the floor plan and priority shut-off list for water, gas, electricity, cable, telephone, alarm and sprinkler systems.

Traffic/Crowd Control Equipment

Expect that parents will want to park as close to the school building as possible without regard to the space needed for fire equipment and vehicles of first responders unless there are barriers to prevent them from doing so. Store traffic cones, caution tape and directional signs. Laminate and store a large map that directs parents to the student release sign-in area.

Sanitation Supplies

If the building is not immediately accessible after an emergency, portable toilets may need to be improvised. Store porta potties (portable toilets), tarps to use as privacy screens, toilet liners, paper and disinfectant, waterless antibacterial soap, water carriers, washing and cleaning supplies. Create directional signs and signs that identify separate temporary toilets for girls and boys. Create, laminate and store these signs in advance.

Water for Storage

After a major disaster, it may take public works several days before the water distribution system can be restored. Even then, the water may require additional disinfection treatment to be safe to drink. Storing an ample emergency supply of clean water should be a top priority.

The shelf life of water depends on its source, how it is treated and how it is stored. Commercially bottled water is the safest and most reliable for storage. According to the Food and Drug Administration, professionally bottled water that complies with legal safety standards and is unopened will remain safe to drink indefinitely. Even if the water develops an unpleasant odor or taste, it should still be safe to drink, according to the FDA.

Tap water, however, must be treated to make it safe to store for six months. There are several methods for treating water: boiling, distillation and chlorination, which uses ordinary household liquid bleach to kill microorganisms. The potency of bleach diminishes with time, so use bleach from an unopened bottle. Follow the recommended proportions, 16 drops of bleach to each gallon of water. The bleach is safe to drink in such small concentration, but avoid using scented bleaches, color-safe bleach and bleach with added cleaners.

Water should be stored in an easily accessible, dark, cool, dry area away from any solvents or chemicals. Water is needed for hygiene and drinking so a multi-strategy approach to storage is recommended. The general rule is for water storage is one gallon per person per day for a three-day supply. Hot environments may need up to double that amount. Store two quarts per person per day for drinking and two more quarts for daily sanitation and food preparation.

A fifty-gallon drum of water is immobile and requires a pump and a distribution system. If water is stored in large drums, it should be treated and replaced every six months. Using this as the only emergency water storage for the school presents the risk of a single point of failure if the water somehow becomes contaminated.

For hygiene, seven or five-gallon containers are preferable. Store a dolly in the emergency cache to transport the water containers. For drinking, schools may prefer to acquire 4 oz. pouches for the youngest students and cases of 8 oz. bottles for those who are older. Store all bottled water in sealed, unbreakable containers, not glass, and don't open them until they are needed. Individual plastic bottles will create more debris.

Food for Storage

It has been said that people will eat anything if they are hungry enough. Kids, not so much. But most of the mealtime foods they enjoy are unsuitable for year-long storage in an emergency container that experiences extremes of heat and cold. Some students may be allergic to nut oils, others to gluten or wheat. Salty items will make them thirsty, and sugar may make them hyperactive.

So give careful thought to what types of food will be stored. The five-year shelf life food bars are a popular choice because they provide complete nutrition and require no refrigeration, preparation, cooking or water. Some bars taste like stale cookies and others are quite oily, so it is advisable to sample them before purchase.

All food should be stored in airtight containers. Keep long-term food storage in cool, low light areas for longest shelf life. Store it off the floor and away from direct contact with any walls in case moisture seeps in. Perishable food should be replaced annually.

Organization is critical when it comes to food storage. Label each food container with the contents and the date they were purchased. Place the oldest items in the front. By purchasing the items in stages and rotating them out annually, the container will never face the problem of having to replace the entire cache.

Shelf Reliance

The shelf life of food depends on the type of food, the packaging and the storage conditions. Light, heat and moisture in the storage container will all affect the shelf life of foods. Exposure to too much light may cause deterioration, particularly affecting fats, oils, proteins and vitamin loss. Heat breaks down the nutritional value of foods and may affect the odor. Excess moisture creates an environment that supports the growth of microorganisms that can cause illness.

Dried milk can be purchased in bulk and stored for up to two years in an airtight container. Low-acid foods last longer than foods with higher amounts of acid including canned tomatoes and fruits. Moisture has been eliminated from dehydrated or freeze-dried foods so they are suitable for long-term storage but may be more expensive. Replace commercially canned food every year.

FEMA and the American Red Cross have published a booklet that includes general guidelines for replacement of common emergency foods. The shelf life of common foods is described in a document titled, "Food and Water in an Emergency" on the FEMA website.

SHELF-LIFE OF FOODS FOR STORAGE

Use within six months:

- Powdered milk - *boxed*
- Dried fruit
- Dry, crisp crackers
- Potatoes

Use within one year, or before the date indicated on the label:

- Canned condensed meat and vegetable soups
- Canned fruits, fruit juices, and vegetables
- Ready-to-eat cereals and uncooked instant cereals
- Peanut butter
- Jelly
- Hard candy and canned nuts
- Vitamins

May be stored indefinitely (in proper containers and conditions):

- Wheat
- Vegetable oils
- Dried corn
- Baking powder
- Soybeans
- Instant coffee, tea, and cocoa
- Salt
- Noncarbonated soft drinks
- White rice
- Bouillon products
- Dry pasta
- Powdered milk – *in nitrogen-packed cans*

Source: Federal Emergency Management Agency and American Red Cross http://www.fema.gov/

First Aid Supplies

Medical supplies for school staff and students should only be purchased in consultation with the school nurse. Many items to be stored are obvious: adhesive tape, bandages in assorted sizes, cold packs and cotton balls, eye wash, face masks, finger splints and backboards, hypo-allergenic tape, hydrogen peroxide solution, vinyl or non-latex gloves, sterile water pouches, saline for washing wounds, sterile gauze pads and rolls, scissors and tweezers, thermometers, tongue depressors, stretchers, waterless anti-bacterial soap and antiseptic hand wipes. Before purchasing restricted use over-the-counter medication (including Tylenol, Ibuprofen, anti-diarrhea cream, Ipecac®, antihistamines and antibiotic cream), check what the school policy is regarding dispensation of these medications to students in an emergency.

Consider the unique needs of the school community in order to create well-stocked medical supply area and remember to include medical documentation: emergency first aid instructions, shock/trauma response instructions, legal responsibility release forms, and triage tags. Store with the medical supplies a list of students and staff with defined allergies and a list of students requiring specified medication, with instructions for dispensing those drugs. Store sanitary napkins both for hygiene and for absorbing blood. Clipboards, marking pens, highlighter, notepads, Post-it® notes and Ziploc® bags are other useful items that should be included with the medical supplies.

Parent Responsibilities

Normal around here is just a setting on the dryer.

Word spread quickly about the local bank heist in the small suburban community. While law enforcement officers scoured neighborhoods for the robber, who fled on foot, the nearby elementary school initiated immediate lockdown procedures. As policemen cordoned off a security zone around the campus to allow them to search the area, parents who heard about the threat of an intruder began arriving to pick up their children and congregated outside the school. Several anxious parents became loud and aggressive, breached the perimeter and banged on school doors and windows. Their pleas went unanswered until law enforcement officers turned them away, reminded them of the potential risks in their actions and subdued their anxieties.

The schools have critical emergency management responsibilities. They are required to have a comprehensive safety plan that includes emergency preparedness, response and recovery and to test the plan annually or biannually. Education Code mandates periodic drills and staff must receive appropriate training.

Parents have responsibilities, too. It is critical that parents and students follow the instructions of school staff and first responders during an emergency. Become familiar with school response procedures so that the safety of others is not compromised. When a lockdown is activated, for example, no one goes in and no one goes out. If an early student release is implemented, parents must remember that only specifically-designated adults can pick up their children; a visiting relative or a neighbor may not have been authorized to do so.

Here are a dozen things parents can do to reduce risks and support school response to emergencies.

1. **Keep your child's emergency card up to date.** The only people other than yourself who will be allowed to pick up your child are those whom you authorize on the Student Emergency Contact Card. No student will be allowed to leave with another person, even a relative, unless the school has prior written permission from the parent/guardian. Notify the school if your child has special needs.

2. **Make and practice a family plan**. Teach your child how to recognize danger signals such as smoke detectors, fire alarms and local community warning signs. Explain how and when to call for help.

3. **Prepare your child.** Children who are prepared experience less fear and hysteria. Let your child know who can make the pickup at school if you are unable to do so. Reassure your child that he/she will be cared for at school until you arrive. Place contact information and phone numbers in your child's backpack.

4. **Help your child memorize important family information:** name, address, phone number, where to meet in an emergency, and who is authorized for pickup from school.

5. **Keep emergency supplies in your car,** including comfortable walking shoes, water and warm jackets and **carry the pocket contact card** in your wallet with phone numbers for your child's pediatrician, the adults you authorized to pick up your child and an out-of-state contact who may be easier to reach if local phone lines are out of service.

6. **Become familiar with the established procedures** and follow directions in an emergency.

7. **Walk from your home**, if possible. Leave the streets free for emergency vehicles. You may get to school faster by foot or bicycle.

8. **Park only in areas designed for parents** if you must drive. Emergency vehicles need adequate room for emergency vehicles to park and turn around. Parent vehicles could impede the ability of first responders to get their equipment to the school site.

9. Bring a photo ID when you pick up your child. Always sign out students before removing them from the campus. Students will only be released to their parents or to an adult designated on the Student's Emergency Contact Card.

10. Remain calm. Your child is probably safer at school in the event of a disaster. School personnel are certified in CPR, First Aid and Emergency Preparedness. In California, school personnel, as public employees, are designated as Disaster Service Workers during a declared disaster and may be required to remain for up to 72 hours to assist with student care and perform other assigned duties.

11. Leave the campus immediately. Socializing with friends may impede the evacuation and student release process.

12. Do not call the school. Use the designated hotline for a recorded message to receive updates on the situation. The school phone lines will be needed for emergency communications.

One more thing every family can do: get involved! Local chapters of the American Red Cross and many communities offer classes in First Aid, CPR, personal preparedness and Community Emergency Response Training (CERT).

Drills and Other Activities

You cannot unscramble an egg.

The construction crew remodeling the school's multipurpose room stopped work for their morning break. A young member of the team lit up his first cigarette for the day. Instantly, an explosion rocked the building, burying two workers under a pile of debris. In an adjacent classroom, a teacher and two dozen sixth-grade students were showered with shards of glass and chunks of sheetrock. The traumatized children were surrounded in rubble and immobilized by fear, but an adrenaline rush propelled the school staff into immediate action as they marshaled the students to evacuate the school. The cause of the explosion was not yet known, but the building suffered severe damage and would need to be structurally assessed and repaired.

This scenario introduced a drill conducted at a local school site. A drill is a coordinated, supervised activity, normally used to address a specific action or function, practice skills or test new procedures. The purpose of a drill is to go through actions appropriate for the scenario or simulation presented to perfect consistent response procedures and efficient use of resources, strengthen relationships with emergency partners and increase confidence among school staff, students and the parent community. Response to the drill scenario should answer many questions. Are all students accounted for? Are there injuries that require immediate medical attention? Do all school personnel know their roles in an emergency? What logistical, operational and planning assistance will be needed? Who should be contacted to secure the safety of the school and evaluate the structural damage to the building? How will the cause of the explosion be determined and by whom? What notification will be given to parents? In addressing these questions, a drill may point out vulnerabilities, gaps and weaknesses in the school's Emergency Management Plan.

The words of Vince Lombardi shine optimism on this potential outcome: "Practice does not make perfect. Only perfect practice makes perfect."

Some states mandate schools to hold shelter-in-place and tornado drills quarterly, extreme weather and bomb threat drills twice a year, and annual bus evacuation and crisis response drills. California Education Code requires monthly fire drills be conducted in every public, private or parochial school in the elementary grades and at least four times during the school year at the intermediate level and an earthquake duck, cover and hold on practice at least once each school quarter in elementary schools and once a semester in secondary schools.

The extent of parent involvement in such drills and other school preparedness activities is driven by the need and explicit encouragement of the school administrators and, of course, the interest and enthusiasm of the volunteers. Since an emergency can occur at any time, parents are likely to be on campus, assisting in classrooms, coordinating other support activities and attending school events. With some forethought, parents can be enlisted to assist with drills in preparation for the real deal. They can serve as runners and distribute forms to be completed for a student release program or monitor the school perimeter to assure students do not leave an evacuation area without authorization. They can be asked to provide parking directions to arriving parents in order to assure space is available for the vehicles of emergency responders.

Parents may even be tasked to intentionally test school procedures by amplifying simulated agitation while waiting to meet up with their children or ignoring perimeter controls. Those who lead the Parent Emergency Preparedness Committee might suggest such roles to the school administrators but must earn their trust by being helpful, following instructions and remaining calm.

Resources for Families

Wild ducks and tomorrow both come without calling.

Emergency/Disaster Information

Schools use varying methods of informing the community of campus activities, including deadlines, drills, trainings, performances and celebrations. The weekly printed infosheet, sent home with the students, is, increasingly, the less preferred approach than electronic mail. Many principals ask parents to check the school website frequently. Others prefer to distribute periodic electronic newsletters by email.

Family emergency preparedness is not a popular, uplifting subject. Few people want to plan for or think about an emergency or disaster until it is in front of them. With so many distractions and obligations in modern life, many parents complain that there simply isn't sufficient time to prepare for hypothetical situations that may never come to pass. So consider any school emergency, no matter how small in scope, as an opportunity for a teachable moment. If an evacuation drill is planned, provide parents with simple reminders of student pick-up procedures. Leverage a school lockdown in the news, even if it is in another community or state, as an opportunity to remind parents what to do and not to do to keep their children out of harm's way in a similar situation.

Throughout the school year, provide tips and resource information for parents to assist them with their home preparedness and familiarize them with strategies for response.

Everyone can benefit from this simple reminder put forward by President John Fitzgerald Kennedy, "The time to fix the roof is when the sun is shining."

In a disaster, local relief may take hours or days, so all families should be prepared to cope until help arrives. There are dozens of simple preparedness tips that can be sent home throughout the school year about how to assemble a Disaster Supplies Kit and develop a Family Disaster Plan.

The American Red Cross recommends three simple steps to prepare:

> BE INFORMED. Learn about local hazards, threats, and available emergency resources.
>
> MAKE A PLAN. Include emergency contact information, evacuation routes and where to reunite if the home is unavailable.
>
> GET A KIT. Assemble 5 to 7-day storage of basics: food, water, first aid supplies, tools, sanitation, and special items (documents, pet items).

Here are a few other preparedness tips. Preset the radio station in the car to one of the local emergency broadcast stations. Post emergency telephone numbers by every landline and add ICE (In Case of Emergency) contact information to every mobile phone. Store water in plastic containers, not milk cartons, which will decompose or glass, which might break. Keep an ABC fire extinguisher on each level of the house, and check the pressure gauges annually to ensure they are fully charged. Store hazardous chemicals (paint thinners, bleach, gasoline) away from sources of heat, and secure them to prevent spills. Put sturdy shoes in the car to replace stylish, but impractical, sling back heels in case an evacuation requires a long walk. Remind kids to cooperate fully with school staff and safety officials in an emergency. Engage in a discussion with them about what to do in various types of emergencies.

Provide some simple hypothetical situations to help with the planning. During the day, parents may be in traffic, at work, shopping, in a crowded public place, high-rise or sporting event when an emergency strikes. Do they have an emergency kit in the car and store another one at work? Where would the family go if the home were unsafe? How have they planned to protect family pets?

Develop some simple preparedness activities that involve the entire family. What ten items would they grab if they only had ten minutes to evacuate? Try a learning quiz. How much water is needed per person per day for emergency storage (one gallon per person per day)? What is the school hotline number for emergency information? Who has been authorized to pick up the child? Set family members out on a scavenger hunt to find emergency supplies in the home–from a can opener to a whistle. Advise them to use what has been found to begin assembling the Family Emergency Kit. Preparedness is everyone's responsibility!

Websites

The Internet hosts an abundance of emergency preparedness resources and activities. Those listed here are primary agencies and organizations that serve the country. As a nod to the author's home state, there are several California-centric links, but readers are encouraged to search for their own state's emergency agency websites.

There are, of course many other informative websites and mobile phone applications that are specialized for a locale or type of emergency. The quality of those resources varies greatly, and those website links may be temporary.

While home pages are rarely moved, the website addresses of secondary resources may change. Check the contents, read the reviews, research the facts before relying on specialized resources.

American Red Cross

http://www.prepare.org
http://www.redcross.org/support/emergency-preparedness

The American Red Cross provides extensive information on disaster management, including preparedness tips for home, family, schools and students. Learn about threats that impact families and how to plan and prepare for pets in a disaster.

California Department of Public Health

http://cdph.ca.gov

The California Department of Public Health (CDPH) is dedicated to the health and well-being of California residents. The health information posted on the website provides links to information for communicable, infectious and chronic diseases and conditions from Africanized honeybees to yellow fever.

California Emergency Management Agency

http://www.calema.ca.gov

The California Emergency Management Agency (CalEMA) website links to laws, regulations, preparedness tips, response information management systems, regional operations and other information regarding California emergency services.

Centers for Disease Control and Prevention

http://emergency.cdc.gov

The Centers for Disease Control and Prevention (CDC), located in Atlanta, is the lead federal agency for preventing and controlling disease, injury and disability, and protecting the health and safety of people at home and abroad. The CDC

Emergency Preparedness and Response website keeps the public informed about wide range of public health emergencies.

Federal Emergency Management Agency Ready Program

http://www.ready.gov

FEMA is an independent federal agency that works in partnership with state and local emergency management agencies, federal agencies and the American Red Cross. FEMA's Ready! website is a national campaign to provide individuals with preparedness strategies and details. The website is available in 13 languages. A component is designed specifically for kids to help them participate in family preparedness activities.

Marin County Schools Emergency Services

http://www.marinschools.org

The Emergency Services website of the Marin County Office of Education provides a comprehensive Model Schools Emergency Management Plan, guidelines for student release, advice and resources for parents and best practices to improve emergency preparedness, response and recovery.

National Weather Service (NWS)

http://www.weather.gov

Read weather forecasts, flood and high surf advisories, storm warnings and other weather outlooks for anywhere in the country by city or zip code.

National Oceanic and Atmospheric Administration

http://www.nhc.noaa.gov/

Site of the National Hurricane Center, with links to cyclone advisories, marine forecasts and top hurricane news of the day.

Readiness and Emergency Management for Schools

http://rems.ed.gov

Readiness and Emergency Management for Schools (REMS) is a grant program administered by the U.S. Department of Education's Office of Safe and Drug-Free Schools since 2003 to help school districts improve and strengthen emergency management plans. The REMS website has assembled a resource repository of best practices from grant recipients. These include checklists, forms, organizational charts, planning guides, templates and other tools for schools.

U.S. Department of Health and Human Services

http://www.phe.gov

Public health and medical emergency support, including legal authorities, daily situation update, declared disasters, food safety, infectious diseases and other threats are found on the PHE.GOV website.

U.S. Geological Survey

http://earthquake.usgs.gov/prepare

The Earthquake Hazards Program of the U.S. Geological Survey provides earthquake safety and preparedness resources, frequently asked questions, publications and fact sheets.

Mobile Phone Applications

American Red Cross: Earthquake

http://www.redcross.org/mobile-apps/earthquake-app

Step-by-step instructions for what to do before, during and after an earthquake.

American Red Cross: First Aid

http://www.redcross.org/mobile-apps/first-aid-app

Simple instructions for handling everyday first aid scenarios.

American Red Cross: Hurricane

http://www.redcross.org/mobile-apps/hurricane-app

Instructions for what to do before, during and after the storm; hurricane-related weather conditions and location of local opened American Red Cross shelters.

American Red Cross: Shelters

http://www.redcross.org/mobile-apps/shelter-finder-app

Addresses and map locations of opened local American Red Cross shelters, including current number of shelter residents and date and time of last update.

Centers for Disease Control and Prevention

http://apps.usa.gov/centers-for-disease-control-and-prevention.shtml

Health and safety information on a variety of topics including emergencies and disasters, diseases and conditions, violence and safety.

Federal Emergency Management Agency

http://www.fema.gov/smartphone-app

Tips on how to stay safe and recover after a disaster, an interactive checklist for storing emergency kit and meeting location information, maps of the FEMA Disaster Recovery Centers and shelter locations.

Flashlights

Flashlight apps turn on the LED light of the smart phone camera for use as a flashlight or a strobe.

In Case of Emergency (ICE)

ICE apps allow you to maintain medical data and emergency contacts for first responders and hospital staff in case of a personal emergency.

Marine Service

http://cell.weather.gov/marine/marine.htm

Marine forecast observations including tsunamis and high seas, hurricanes, offshore and local conditions from the National Oceanic and Atmospheric Administration's (NOAA) National Weather Service.

National Weather Service

http://apps.usa.gov/national-weather-service.shtml

Weather data, climate forecasts and warnings for the United States, its territories, adjacent waters and ocean areas from the National Oceanic and Atmospheric Administration (NOAA).

Nation's Health

http://thenationshealth.aphapublications.org/site/misc/app.xhtml

Posts latest health news and lists of public health professionals and decision-makers.

U.S. Geological Survey: Seismic

https://itunes.apple.com/us/app/usgsseismic/id333208233?mt=8

Provides data on recent earthquakes around the world from the United States Geological Survey (USGS). It displays the location, date and magnitude of each earthquake, along with a color-coded graphic that indicates the severity of the earthquake.

Web MD

http://www.webmd.com/webmdapp

Provides access to first aid essentials, symptom checker and closest hospital, physician and pharmacy based on your current location or zip code.

Fundraising

All easy problems have already been solved.

Many schools have insufficient funds to provide all the educational and support services that students need. In too many cases, there is a gap in storing sufficient emergency equipment and supplies. PTAs and parent clubs can get involved in the safety of their children and schools by making a commitment to stock, replenish and enhance the emergency cache of a school site to proactively minimize vulnerability.

Some of the parent-led support activities require little investment of time or money, just a conviction that emergency empowerment is both a necessary and realistic goal. However, acquiring supplies for the classroom or storage container involves unavoidable costs. How can such a project be funded?

First, it's best to begin with realistic goals in terms of funding needs. Develop a three-year plan so that all the startup costs are distributed over time. Don't be afraid to approach the school, which may, indeed, earmark funds for such purpose and appreciate the volunteer assistance with their purchase and organization.

Some schools request a contribution from each family of a nominal amount to contribute to the purchase of supplies. A school with an enrollment of 300 will be well on its way if each family contributes only a few dollars for this purpose.

Parent groups are usually prepared to host fundraising events for a variety of student-centered events but may not be aware of a need to provide support for emergency preparedness supplies. Approach the Executive Committee and ask for funds to be allocated for this purpose annually. Such a program provides an important opportunity to contribute to the safety of their children and their schools.

Most schools have a Back-to-School Night to orient parents and caregivers about school events, services, requirements and personnel. A way of jump-starting fundraising support to get the school prepared for emergency situations is to offer an "Adopt a Classroom" program. Ask teachers to supply a wish list of needed supplies to augment those already on hand. Invite parents to fill the need by individual contributions, group support, or direct purchase of the items.

Another example of the "Adopt a Classroom" program is to accept contributions toward the purchase of one item that every classroom needs, a lockdown kit. Every year, there are media reports of incidents involving an intruder or shooter on campus. In a school lockdown, students and staff are confined to classrooms or other rooms. For the safety of students and staff, a lockdown is initiated and students will remain in the locked classrooms until the threat is removed. That may take several hours.

The lockdown kit provides all the needed supplies for a classroom in a lockdown situation. It includes items such as food and water with a five-year shelf life, radio with batteries, porta-potty, sanitation supplies, duct tape and a solar blanket to build a privacy screen. Check the Appendix for an easy form to contribute to a lockdown kit as a family or a group. Every classroom should have one, and the school will appreciate the support

Fundraising efforts can reach out to the community. Many hardware stores are willing to offer discounts to schools for purchase of emergency supplies. A local flagship corporation or business may have funds available to contribute as part of a philanthropic resilient community program. You won't know unless you ask.

Maintaining Momentum

A rolling stone gathers momentum.

Parent volunteer programs are often difficult to initiate and even harder to sustain from year to year. As children move through the educational structure, from elementary grades to middle school and on to high school, they will change schools and even school districts. Their parent support naturally moves with them.

One passionate parent can provide strong leadership for the emergency preparedness program at a school, and then leave it bereft of support and continuity when his or her child graduates to another school. Planning to keep the effort sustainable should be incorporated into the consciousness of the team members. They should actively recruit their successors and seek membership from parents of children in the younger grades to assure continuity of the work.

PTA organizations and parent clubs publish volunteer opportunities in the spring of the school year and solicit signups for the following fall. Here is a sample description for a Parent Emergency Preparedness Committee Chairperson:

Job Description: Parent Emergency Preparedness Committee Chairperson

READ the district or school Emergency Preparedness Plan to become familiar with its policies and procedures.

ENLIST site coordinators for each school in the district.

COORDINATE support activities with the district or school Risk Assessment Manager, site council and PTA or Parent Club Executive Committee.

MANAGE replacement and replenishment of emergency supplies at each school.

PLAN periodic family emergency preparedness presentations for the parent community.

ADMINISTER funding and maintain budget for emergency preparedness expenses.

Time commitment: Mainly in fall, flexible hours, monthly meetings.

Job Description: Parent Emergency Preparedness Committee Member

ESTABLISH parent telephone tree for the school year and distribute numbers to those who must make calls.

INVENTORY supplies (batteries, first aid, class lists) in each Grab-n-Go classroom backpack. Evaluate and replenish supplies as needed.

REPLENISH the emergency supply cache after it has been used or annually. Replace food and water cache as needed.

EVALUATE supplies and continue to enhance the cache.

COORDINATE the Comfort Kit program.

Time commitment: Mainly in fall, flexible hours, monthly meetings.

Procedures, templates, inventories, vendor lists and other documents should be passed on so that no team has to start anew with the work. Before school closes for the summer, plans should be in place for orientation of the new members who will pick up the gauntlet of emergency support for the school in the fall.

Ten Tips for Sustaining Parent-Supported Emergency Preparedness Activities in Schools

1. ESTABLISH a standing committee for emergency preparedness volunteers within the PTA/parent club support organization.
2. WORK closely with school administration to outline activities that the Parent Emergency Preparedness Committee will focus on each year.
3. MAINTAIN a visible presence at the school through reports to members of the PTA or parent club at meetings.
4. PROVIDE emergency preparedness tips and articles for newsletters and the school website and informational workshops and presentations for the parent community.
5. RECRUIT volunteers from all grade levels, assuring that younger grades are represented to provide continuity of the committee.
6. FORM a committee at the district level, comprised of Parent Emergency Preparedness Committee chairs from every school within the district. Meet monthly to share information, provide mutual support and work toward standardizing procedures within the district.
7. SEEK funding support from PTA/parent club and school sources. Many emergency preparedness committees are funded solely by the PTA/parent support organization.

8. ARCHIVE information about procedures and activities each year. This will provide an orientation for new committee members and document the accomplishments each year.

9. ENJOY the collaboration. Make planning fun. A day without sunshine is like night.

10. ACKNOWLEDGE the work of parent volunteers. This support often gets little attention but it has an important impact on the school community.

Disaster Mythbusters

The nice thing about standards is that there are so many to choose from.

Myths and misconceptions about disasters abound. Some assumptions are perpetuated through exaggerated media reports and others by lack of adequate public education. Just a little misinformation can do a lot of damage. These myths can divert individuals from taking appropriate steps to become informed, make a plan and assemble emergency supplies. Here are ten commonly held misconceptions to dispel in order to reduce risks in the home.

1. It won't happen here. Disasters are rare events.

Although some areas of the country are more prone to certain types of natural disasters–earthquakes in California, tornadoes in the Midwest, hurricanes in Florida, flooding along the Mississippi, blizzards in the Northwest–no area is completely immune from disasters. They are a part of natural life and in many cases are predictable, repetitive events.

2. It won't happen to me. Disasters always happen to someone else.

The personal invulnerability point of view often leads people into an unrealistic and impractical sense of security, believing that they are in some way immune from disasters. No one is exempt from the effects of a disaster.

3. Help is on the way.

Everyone wants to believe there is a white knight who will come to the rescue–firefighter, police officer, neighbor, friend, emergency medical team, American Red Cross. A 2009 survey by FEMA revealed that nearly one-third of U.S. residents don't prepare for disasters because they believe that emergency responders will help them. If the disaster is widespread, the help you are expecting will most likely be busy elsewhere. There simply are not enough first responders in any community to address the needs of every household. As a responsible citizen, it's best to prepare your family so that outside help is not needed.

4. Our house has everything we might need in an emergency.

There is also the opposite misconception that disasters are tricky; it's hard to know what we will need until we need it. Your house has water, food, blankets, warm clothing, flashlights and batteries. If the power goes out, will you be able to find all these items safely? What happens if it is dangerous to go inside your home or you cannot return due to gridlock?

5. Our child is safer with family members than at school.

Teachers and school staff are trained to respond to emergencies of all stripes–earthquakes, intruders, chemical spills, bomb threats, infectious disease, food spoilage, floods and many more. There are plans and procedures in place for all hazards and the first priority of the school is the safety of all students and staff.

6. The insurance company will pay for the damage.

Insurance may not cover everything and many households are underinsured. Many insurance policies exempt "Acts of God," which are natural catastrophes that no one can prevent such as earthquakes, tornadoes, hurricanes and tidal surges. Some insurance companies are redefining exclusions to incorporate terrorism. Coverage of these hazards and others often requires a separate policy for each. Become familiar with the terms of your homeowner's insurance policy.

7. It costs too much to prepare.

There are many simple and economical ways to begin family preparedness and drastically improve an ability to deal with an emergency.

8. Disasters are uncontrollable events that cannot be managed.

First responders are vigorously trained to understand how disasters function and are managed. A common language and functional response structure brings efficiency and effectiveness to disaster management.

9. The less said to children about a tragedy or disaster, the better.

Everyone wants to know what is happening when significant events are in flux. Youth are no exception. Restricting access to information creates a lack of trust and leaves an opportunity for rumors to fill the void. Information should be shared in a calm and age-appropriate manner. It is important to talk to children before any emergencies occur and discuss how you will respond as a family. Children also need to be taught to obey instructions during an emergency and respond quickly to evacuation orders. Descriptions of worst-case scenarios are not helpful. Instead, assurance that there are plans in place, and that school staff and responders know what to do, will help allay fears.

10. Everything will be back to normal within a few weeks.

This is wishful thinking. Recovery from a disaster can take years. The physical, financial and emotional effects of a disaster last a long time and people may never be able to fully repair their lives if the infrastructure of the community is disrupted, businesses are shuttered, jobs are terminated and financial resources are depleted.

Frequently Asked Questions

Where would we be without rhetorical questions?

The process of disaster planning requires developing scenarios and then establishing procedures, policies and assignments to meet the varied requirements of each one. Even with an emergency management plan in place, many questions remain.

1. Why don't middle school students have comfort kits?

Students of this age are old enough to understand that parents may be delayed in an emergency. Parents should remind them, in advance, to follow the instructions of the school staff and reassure them that they will be picked up as soon as possible.

2. What can we do about mildew in the emergency supply boxes and containers?

Mildew and mold are caused by excess moisture that is trapped and by poor ventilation. Moisture can also wick up into fabric placed on the container floor. Check the seams of the container to find the source of the moisture. Install ventilation fans to reduce the amount of moisture. Store equipment supplies in a dry container. Use plastic, not cardboard, boxes with tight-fitting lids. Use an all-purpose cleaner to scrub walls and floors. When the surfaces are dry, mop them with a mixture of one cup of bleach to one gallon of water. If fabrics are safe for chlorine bleach, wash the in hot water, laundry detergent and the amount of bleach recommended on the label.

3. If an immediate evacuation is required, what should students take from the classroom?

Evacuations generally require fast response. This is not the time for students to be collecting their books, backpacks or lunchboxes.

If the class is unable to return to the room after a short duration, the school will provide water and quiet entertainment. Comfort kits may be distributed to students in the lower grades to occupy them until their parents arrive.

4. What if parents can't get to school to pick up their children?

In the event of a major emergency, it is very likely that some parents may be unable to reach the school. Students will be released to any adult who is listed on the emergency contact card as authorized to pick them up. A written record should be kept by the school that records when the student was released and to whom.

5. Where is the best place to store emergency supplies in the home?

The best place to store the family emergency supplies is not in the home or the garage, but outside. Choose a location that is not in a gully where water can collect, not below trees with limbs that can break or power lines that can fall. Use a waterproof container with a tight-fitting lid and fasten with a bungee cord as extra protection to prevent raccoons and other wildlife from breaking into the supplies.

6. Is there a time frame on when products in the emergency kit without an expiration date should be used?

Check the list in the chapter on storage and supplies for the shelf life of foods. Products that have no expiration date should not be used if there are signs of discoloration, degradation or damage. Those that sit in the container over the summer months should be replaced in the fall. When in doubt, throw it out.

7. Will ATMs work in a disaster?

ATM machines rely on electricity. When the power is out, the ATM machines, gas pumps and cordless telephones won't work. Store cash and small change in your car and home emergency kit. Fill up the tank when it reaches the halfway mark.

8. Will cell phones work in a disaster?

Mobile phone communication can play an important role in helping recover from a disaster. However, the transmitters are oversubscribed, and in the immediate aftermath of a disaster, cell phone callers will most likely experience congestion. Text messages are more likely to get through because texts use less bandwidth and can easily be queued and sent when capacity is available. The key message is *text not talk* during emergencies.

9. If the school is evacuated, how will parents know where to find their children?

If relocation is required, information will be posted at the school. In addition, many schools have automated telephone notification systems that will relay a recorded message quickly to the parent community. Some of these systems also allow for email alerts and text messages to cell phone numbers. Most schools have a dedicated telephone that serves as a hotline for recorded updates or a phone number assigned for that purpose during an emergency.

10. How much information should we share with our children about a large-scale emergency or disaster?

It bears repeating. Restricting access to information creates a lack of trust and leaves an opportunity for rumors to fill the void. Information should be shared in a calm and age-appropriate manner. It is important to talk to children before any emergencies occur and discuss how you will respond as a family. Children also need to be taught to obey instructions during an emergency and to respond quickly to evacuation orders. Descriptions of worst-case scenarios are not helpful. Instead, assurance that there are plans in place, and that school staff and responders know what to do, will help dispel their fears.

Appendices

The beginning of wisdom is to call things by their right names.

A. Acronyms

Why use words when an alphabet soup of sometimes thoughtful, often irreverent acronyms can convey the message in more succinct form? Here is an opportunity to apply a little levity to emergency preparedness activities. Forget LOL and OMG. Here are AMFYOYO, JASA, IIS, and BWNDITW for those of the texting persuasion. YMMV!

AFAIK As Far As I Know
AFAP As Far As Possible
AMFYOYO Adios, My Friend, You're On Your Own
AOG Act Of God
ARC American Red Cross

BBIAB	Be Back In A Bit
BC	Before Computers
BLS	Bring Long Splints
BION	Believe It Or Not
BWNDITW	But We've Never Done It That Way
BRB	Be Right Back
CalEMA	California Emergency Management Agency
CERT	Community Emergency Response Team
CDC	Centers for Disease Control and Prevention
CHOP	Choking On My Popcorn
DCAP	Disheartened Crazy Ambulance Paramedic
DWS	Driving While Stupid
EMT	Emergency Medical Technician
EOC	Emergency Operations Center
EOP	Emergency Operations Plan
ETA	Estimated Time of Arrival
FAQ	Frequently Asked Questions
FEMA	Federal Emergency Management Agency
FIFI	First In, Forgotten Immediately
FILO	First In, Last Out
FINO	First In, Never Out
FISH	First In, Stays Here
FITB	Fill In the Blank
FUBB	Fouled Up Beyond Belief
GUP	Grace Under Pressure
HAZMAT	Hazardous Materials
IIS	Impossible to Initiate Safely
IMHO	In My Humble Opinion
JABBA	Just Another Brain Boggling Acronym
JIT	Just In Time
LERG	Let Everyone Rage and Groan
MOPAR	Mostly Old Parts and Rust
MRE	Meals Ready to Eat
MTTR	Mean Time To Repair
NIMS	National Incident Management System
NINO	Nothing In, Nothing Out
NTF	No Trouble Found
NWS	National Weather Service
OES	Office of Emergency Services

OMG	Oh My God
OOTB	Out Of The Box
PIO	Public Information Officer
PPE	Personal Protective Equipment
PTB	Powers That Be
SEMS	Standardized Emergency Management System
SFSG	So Far, So Good
SNMP	Simply Not My Problem
TARP	Telephone Access Routing Plan
TBD	To Be Determined
TMI	Too Much Information
TYT	Take Your Time
USGS	U.S. Geological Survey
VOAD	Voluntary Organizations Active in Disaster
WIIFM	What's In It For Me?
WIP	Work In Progress
WOMBAT	Waste Of Money, Brains And Time
WPF	When Pigs Fly
YABA	Yet Another Bloody Acronym
YMMV	Your Mileage May Vary
YTYTK	You're Too Young To Know

B. Emergency Supplies Checklist

FEMA's Emergency Supply List is available for download at: http://www.ready.gov/. Readiness guidance has been translated into 12 languages, in addition to English.

The American Red Cross has published a list of basic supplies at: http://www.redcross.org/prepare/location/home-family/get-kit.

The 72 Hours website at http://www.72hours.org, produced by the San Francisco Department of Emergency Management, provides guidance in English and four additional languages–Spanish, Chinese, Russian and Vietnamese–for building a kit of emergency supplies and responding to a multitude of emergency scenarios from contagious disease to terrorism. The website also lists preparedness guidance for children, seniors and individuals with special needs.

The California Emergency Management Agency has developed a succinct, single-page Emergency Supplies Checklist, shown on the next page.

PREPAREDNESS TIPS

Cal EMA
CALIFORNIA EMERGENCY MANAGEMENT AGENCY

Emergency Supplies Checklist

The first 72 hours after a major emergency or disaster are critical. Electricity, gas, water, and telephones may not be working. In addition, public safety services such as police and fire departments will be busy handling serious crises. You should be prepared to be self-sufficient - able to live without running water, electricity and/or gas, and telephones - for at least three days following a major emergency. To do so, keep the following on hand in a central location:

Can You Go it Alone for Three Days?

Essentials

- ☐ Water - One gallon per person per day (a week's supply of water is preferable)
- ☐ Water purification kit
- ☐ First aid kit, freshly stocked
- ☐ First aid book
- ☐ Food
- ☐ Can opener (non-electric)
- ☐ Blankets or sleeping bags
- ☐ Portable radio, flashlight, and spare batteries
- ☐ Essential medications
- ☐ Extra pair of eyeglasses
- ☐ Extra set of house and car keys
- ☐ Fire extinguisher - A-B-C type
- ☐ Food, water, and restraint (leash or carrier) for pets
- ☐ Cash and change
- ☐ Baby supplies: formula, bottle, pacifier, soap and baby powder, clothing, blankets, baby wipes, disposable diapers, canned food and juices

Sanitation Supplies

- ☐ Large plastic trash bags for waste; tarps and rain ponchos
- ☐ Large trash cans
- ☐ Bar soap and liquid detergent
- ☐ Shampoo
- ☐ Toothpaste and toothbrushes
- ☐ Feminine hygiene supplies
- ☐ Toilet paper
- ☐ Household bleach

Safety and Comfort

- ☐ Sturdy shoes
- ☐ Heavy gloves for clearing debris
- ☐ Candles and matches
- ☐ Light sticks
- ☐ Changes of clothing
- ☐ Knife or razor blades
- ☐ Garden hose for siphoning and firefighting
- ☐ Tent
- ☐ Communication kit: paper, pens, stamps

Cooking

- ☐ Plastic knives, forks, spoons
- ☐ Paper plates and cups
- ☐ Paper towels
- ☐ Heavy-duty aluminum foil
- ☐ Camping stove for outdoor cooking (caution: before using fire to cook, make sure there are no gas leaks; never use charcoal indoors)

Tools and Supplies

- ☐ Axe, shovel, broom
- ☐ Adjustable wrench for turning off gas
- ☐ Tool kit including a screwdriver, pliers and a hammer
- ☐ Coil of 1/2" rope
- ☐ Plastic tape, staple gun, and sheeting for window replacement
- ☐ Bicycle
- ☐ City map

For more info, visit www.calema.ca.gov

Source: California Emergency Management Agency
http://www.calema.ca.gov

C. Supplies Replacement Form

EMERGENCY CACHE – SUPPLIES NEEDED					
ITEM - Identify M=Missing, R=Replacement	QTY	REQUESTED BY	DATE	VENDOR if known	COMMENTS

D. Classroom Lockdown Kit

The Adopt a Classroom Lockdown Kit is a concept developed and distributed by Earthquake Management in Westminster, California. The Classroom Lockdown Kit is shown below in a flyer reprinted with permission from Earthquake Management.

YOUR SCHOOL NAME HERE

Adopt A Classroom

While it is true that our schools are among the safest places our children can be, we are still affected by the violence that does exist in our world today. Because of this, we practice lockdown drills regularly and are continually improving our procedures to prevent violence at school and to keep our students & staff safe.

We are inviting parents, friends, and community members to help us meet the needs of students during lockdowns by **Adopting a Classroom** by providing a lockdown kit for a classroom. Each kit contains all the needed supplies for a classroom in a lockdown situation. **Even $2.00 donations will help.**

3 3600 calorie food bars (27 meals)
30 packs of water
1 portable toilet
100 handy wipes
Toilet Paper
12 Toilet Liners
2 Toilet Chemicals
Radio / flashlight
1 whistle
1 Roll of duct tape
1 solar blanket (to use as a privacy shelter or blanket w/ duct tape)

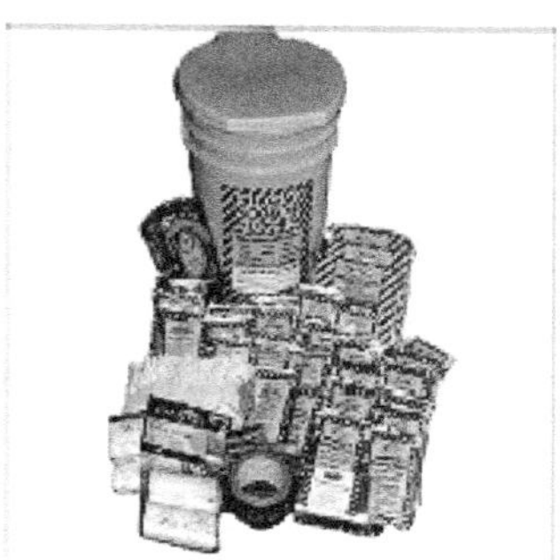

Mayday food & water Have a 5 year shelf life and are U.S. Coast Guard approved

Mayday Food Bar:
100% of all essential vitamins.
Safe for vegetarians or anyone allergic to nut oils.
Non-thirst provoking
Expands in stomach
No Cholesterol
Will tolerate up to 149 degrees temperature

- - - - - - - - - - - - - - - *please fill out and return this portion* - - - - - - - - - - - - - - - -

Yes! I would like to adopt a classroom.
School: ____________ Name: ____________
Classroom: ____________ Address: ____________
Attach a check for $______ Phone: ____________
E-mail address: ____________________
I am interested in purchasing kit for home: ____

Source: Earthquake Management
http://www.earthquake-management.com/kits.html

About the Author

Ellen Rony has worked on emergency preparedness activities since 1993, when she led the development of a program for the Reed Union School District of Tiburon, California. She established a PTA Emergency Preparedness Committee that acquired and organized a supply cache for three school sites, designed a phone tree for the parent community and launched a comfort kit program for all the elementary school students. The program has been showcased as a model for other school districts.

In 1997, she developed an Emergency Services website for the Marin County Office of Education to provide guidelines and plans for school administrators, resources for parents, emergency contacts, supply lists, training schedules, links to community agencies and other materials. She continues to serve as its webmaster.

Ellen wrote a federal grant application to the U.S. Department of Education and U.S. Department of Homeland Security in 2003 on behalf of the Marin County Office of Education to strengthen preparedness in local schools. She was hired to coordinate the grant activities for the next two years. During that period, she launched the Parents Disaster Advisory Council, wrote a comprehensive Model Marin Schools Emergency Management Plan and led the development of guidelines for emergency student release.

In 2009, Ellen co-authored another U.S. Department of Education application for a Readiness and Emergency Management for Schools grant from the U.S. Department of Education. She served as the grant's project coordinator for the Marin County Office of Education through 2012.

Ellen and her brother, Peter, co-authored *The Domain Name Handbook: High Stakes and Strategies in Cyberspace*, published in 1998. She developed a companion website to the book at

DOMAINHANDBOOK.COM and a website to honor the film legacy of their father, George Rony, at GEORGERONY.COM

Questions for the author should be addressed by email to: erony@alumni.stanford.edu.

Be prepared!

Made in the USA
Las Vegas, NV
28 July 2024

93028892R00049